ANIMALS IN MILITARY MEDICINE

by Amy C. Rea

abdobooks.com

Published by Pop!, a division of ABDO, PO Box 398166, Minneapolis, Minnesota 55439.

Printed in the United States of America, North Mankato, Minnesota.

102021
012022

THIS BOOK CONTAINS RECYCLED MATERIALS

Cover Photo: Shutterstock Images
Interior Photos: Shutterstock Images, 1, 26; Defense Visual Information Distribution Service, 5, 6, 7, 14, 16 (top), 17, 19, 21, 29; 615 collection/Alamy, 8; Chronicle/Alamy, 11; Historic Collection/Alamy, 12; iStockphoto, 13, 22, 23 (Australian shepherd), 23 (border collie), 23 (boxer), 23 (Doberman pinscher), 23 (German shepherd), 23 (golden retriever), 23 (Labrador retriever), 25, 28; Department of Defense, 15; Sputnik/Alamy, 16 (bottom)

Editor: Charly Haley
Series Designer: Laura Graphenteen

Library of Congress Control Number: 2020948929

Publisher's Cataloging-in-Publication Data

Names: Rea, Amy C., author.
Title: Animals in military medicine / by Amy C. Rea
Description: Minneapolis, Minnesota : Pop!, 2022 | Series: Military animals | Includes online resources and index.
Identifiers: ISBN 9781532169946 (lib. bdg.) | ISBN 9781644945896 (pbk.) | ISBN 9781098240875 (ebook)
Subjects: LCSH: Animals--Juvenile literature. | Working animals--Juvenile literature. | Medicine, Military--Juvenile literature. | Armed Forces--Juvenile literature.
Classification: DDC 355.424--dc23

WELCOME TO DiscoverRoo!

Pop open this book and you'll find QR codes loaded with information, so you can learn even more!

Scan this code* and others like it while you read, or visit the website below to make this book pop!

popbooksonline.com/medicine

*Scanning QR codes requires a web-enabled smart device with a QR code reader app and a camera.

TABLE OF CONTENTS

CHAPTER 1

READY FOR DISASTER

Broken pieces of rock and metal sit on the ground. Chyta sees them. She knows she has a job to do. She rushes back and forth across the rubble. Her nose is close to the ground. She climbs over

WATCH A VIDEO HERE!

Search and rescue dogs are trained to find people who may be buried under rocks after a disaster.

the broken rock. This is only for practice. Still, **drills** are important. They help Chyta get ready for real disasters. Chyta is a search and rescue dog.

People have trained search and rescue dogs for hundreds of years. Militaries began using search and rescue dogs during World War I (1914–1918).

After a search and rescue dog finds someone, that person can get medical care.

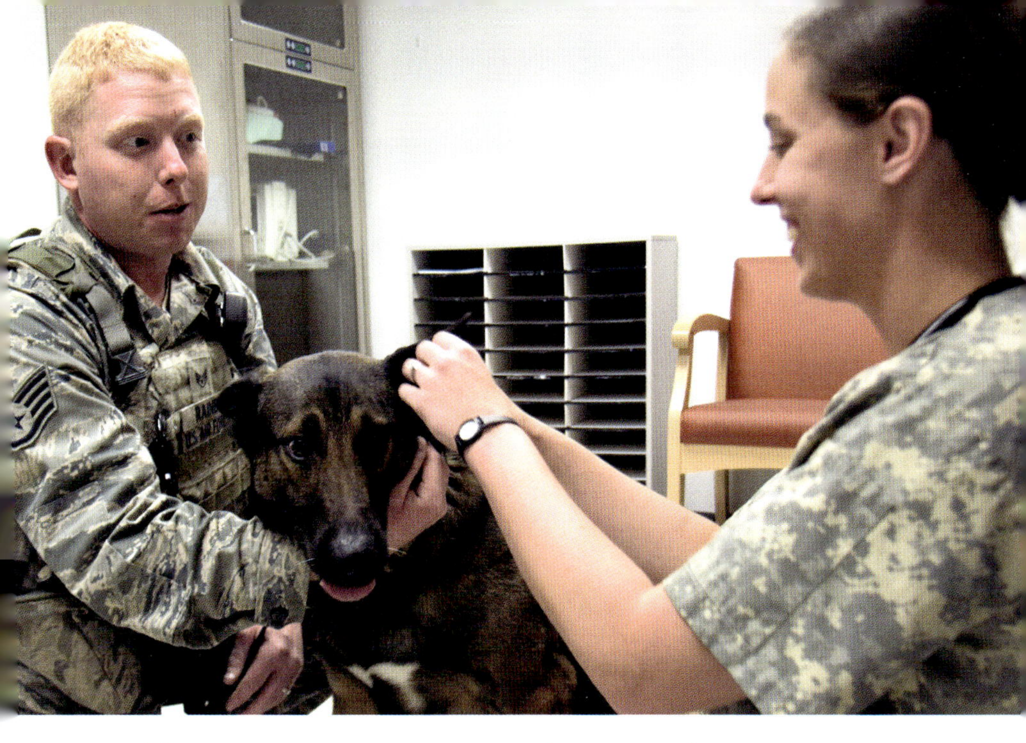

Military dogs help take care of hurt soldiers, but the dogs need care too. Military veterinarians make sure dogs are healthy.

These dogs found hurt soldiers.

Today, search and rescue dogs find people who have been hurt or trapped in disasters. Militaries also use these dogs to find survivors after bombs go off.

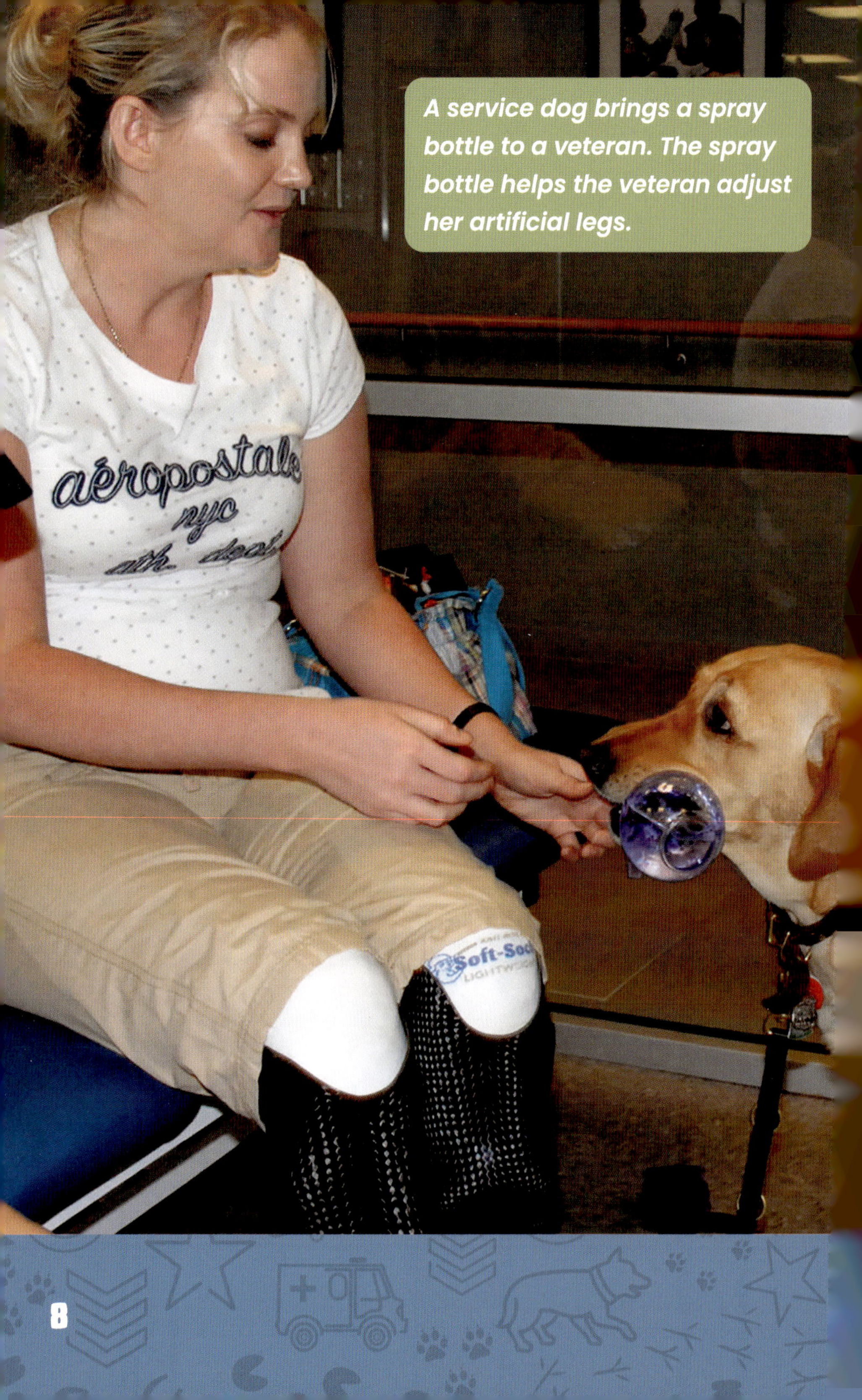

A service dog brings a spray bottle to a veteran. The spray bottle helps the veteran adjust her artificial legs.

Many animals help militaries. For example, there are search and rescue dogs like Chyta. There are also service dogs. Those dogs help **veterans** who have **disabilities**. Other kinds of animals help military doctors treat soldiers. All of these animals are important to members of the military.

People should not pet service dogs without permission. The dogs need to do their jobs. They cannot be distracted.

CHAPTER 2

MEDICAL ANIMAL HISTORY

Militaries have used medical animals for many years. Mules and horses used to pull ambulances. They carried hurt soldiers to safety during World War I. Camels carried hurt soldiers too. Militaries used camels in deserts.

LEARN MORE HERE!

British nurses used horse-drawn ambulances during World War I.

DID YOU KNOW?

In World War I, medicine dogs carried first aid kits to wounded soldiers on the battlefield.

Other animals had different jobs. For instance, cats and dogs killed mice and rats in the **trenches** during World War I. Mice and rats can carry diseases. Killing them helped the soldiers stay healthy.

A soldier gets bandages from a medicine dog during World War I.

German shepherds are strong and loyal. These traits make them good service dogs.

Military guide dogs were first used in Germany during World War I. A doctor in a **veterans'** hospital saw a German shepherd leading someone who was blind around safely. The doctor decided to start training dogs to serve veterans.

Many soldiers see awful things during wars. They are often in danger. These situations cause stress. Animals can offer them comfort. Soldiers have adopted animals for this reason throughout history. The US Army began using **therapy** dogs in 2007. The dogs help soldiers handle the stress of being in battle.

A therapy dog licks a soldier's face.

US military members pet a therapy dog during the COVID-19 pandemic in 2020.

TIMELINE

1600s

People train search and rescue dogs to find lost travelers in the Swiss Alps.

1750s

The world's first guide dogs are trained in Paris, France.

1914–1918

Mules and horses pull ambulances during World War I.

2007

The US Army begins using animals for therapy. Two dogs are sent to soldiers in Iraq.

1914–1918

During World War I, cats and dogs kill mice for soldiers.

2020

US lawmakers work to start a program that would help pay for more veterans to get therapy dogs.

CHAPTER 3

SERVICE AND THERAPY ANIMALS

Service dogs help **veterans** who were hurt in **combat**. A service dog can help a person who is blind move around safely. Dogs can be trained to get things for

COMPLETE AN ACTIVITY HERE!

Service dogs can help veterans who use wheelchairs.

people who have lost the use of their arms or legs. These dogs can also guide people up and down stairs.

Service dogs may learn to turn on light switches or get things from a refrigerator.

SERVICE AND THERAPY DOGS

Service dogs and therapy dogs both help people in the military. But they are different. Service dogs are trained to do work. They learn many tasks to help people with disabilities. They must be with their person at all times. That is why service dogs are allowed into buildings that do not normally allow pets. Therapy dogs comfort people. They do not have to learn the tasks that service dogs learn. They are not allowed in all buildings.

Service and **therapy** dogs both help soldiers who have seen combat. The dogs help ease their stress.

Therapy dogs do not need special training. But they must pass behavior tests. They have to be calm at all times.

Military animals often wear vests to show people they are on the job.

Service dog training can take up to two years. The dog must learn basic commands such as "sit" and "stay." The animal learns how to stop at street curbs and walk in a straight line. These behaviors help the dog guide people.

A service dog trains to help people who are blind walk around obstacles.

Service dogs must also learn to stay calm. They learn to work in busy places. They must be able to handle crowded streets and lots of noise.

SERVICE AND THERAPY DOG BREEDS

Australian shepherd

Doberman pinscher

Boxer

German shepherd

Border collie

Golden retriever

Labrador retriever

*Several dog **breeds** are commonly trained to become service and therapy animals.*

CHAPTER 4

OTHER MEDICAL ANIMALS

Animals such as pigs, mice, goats, and dogs have all been part of military medical research. Military doctors use these animals to learn how to treat injuries. Practicing on animals is more

LEARN MORE HERE!

realistic than practicing on **mannequins**. Some people do not agree with this kind of research. They believe it is harmful to the animals.

Scientists use mice to test new kinds of medicine.

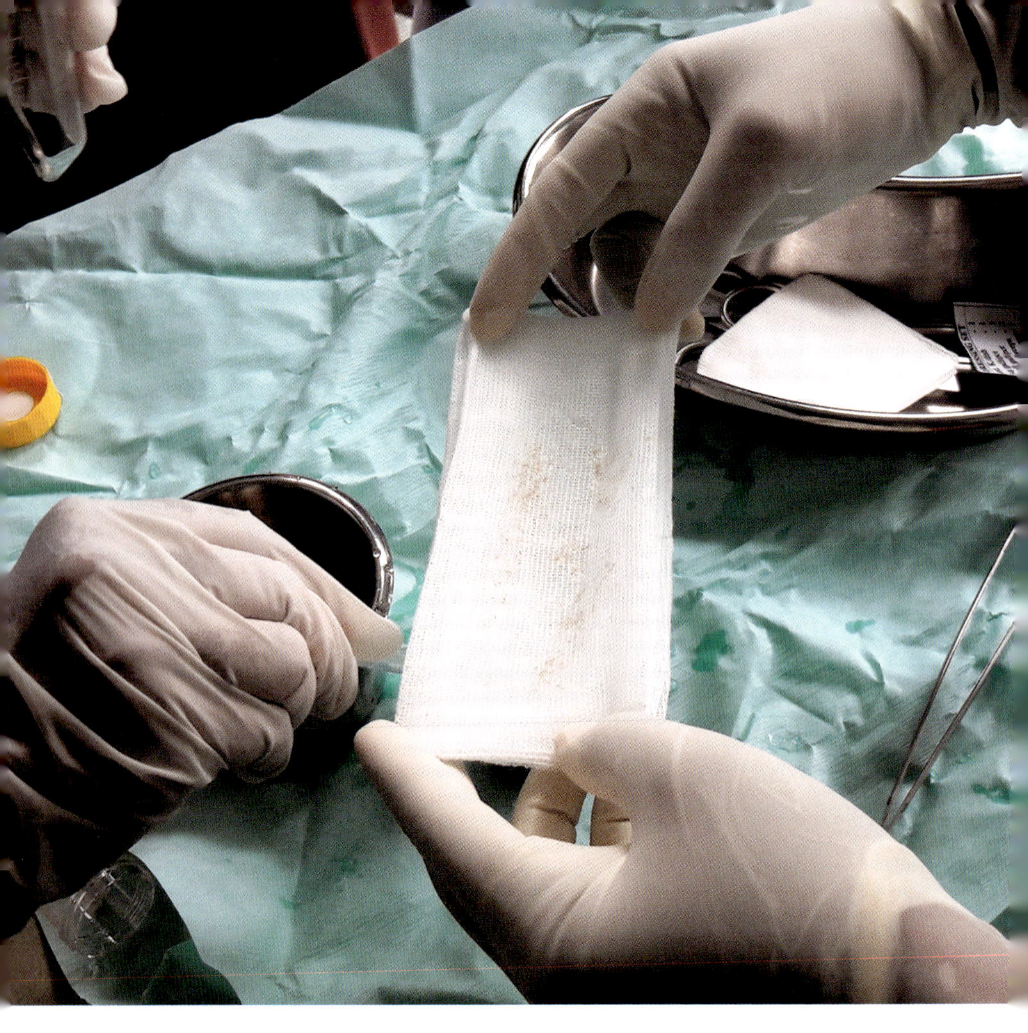

Doctors in Malaysia prepare maggots to help patients.

Treating soldiers in the field can be tricky. Doctors must keep wounds clean. Animals can sometimes help with this.

Maggots are wormlike animals. Military doctors can place maggots on a wound. The maggots eat the dead skin. They do not eat living skin. In this way, maggots help injuries heal more quickly. Militaries around the world can use maggots in emergencies.

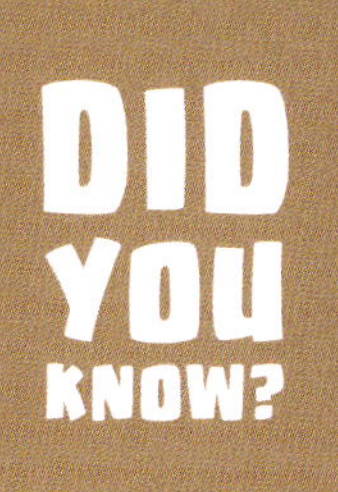

In 2019, the United Kingdom was working to send maggots to help treat people injured by war in Syria, Yemen, and South Sudan.

Military animals help in many areas of medicine. Some work in hospitals. **Therapy** dogs and cats comfort people. Others work in the field. They provide both treatment and research.

Therapy dogs are calm, and they offer comfort to soldiers.

US military dogs help people from all areas of the military.

MAKING CONNECTIONS

TEXT-TO-SELF

Have you ever helped train an animal? If so, what was it like? If not, how would you train an animal to do what you want?

TEXT-TO-TEXT

Have you read any other books about animals in the military? How were those animals similar to or different from the animals in this book?

TEXT-TO-WORLD

Have you ever seen a service dog or other type of animal working in real life? What was the animal doing?

GLOSSARY

breed — a specific type of animal.

combat — fighting between enemies in war.

disability — a condition that limits a person's physical or mental abilities.

drill — an activity to practice and perfect certain skills.

mannequin — a fake human body made out of plastic or other material.

therapy — a form of medical treatment, often for mental health.

trenches — deep, narrow pits that many soldiers fought from and hid in during World War I.

veteran — a former member of the military.

INDEX

ONLINE RESOURCES

popbooksonline.com

Scan this code* and others like it while you read, or visit the website below to make this book pop!

popbooksonline.com/medicine

*Scanning QR codes requires a web-enabled smart device with a QR code reader app and a camera.